"Amazing" "Beautiful" "Historic"

Apollo Launch Photos From the Earth to the Moon and back again

by Ed Bernd Jr.

More information

For more information and free pictures to download, and if you'd like to buy photographic-quality prints with an option for custom framing, please visit www.ApolloLaunchPhotos.com

Acknowledgements:

Special thanks to Jeanetta Wright, Wes Moore, Jose Escalera, and Katherine Sandusky for their encouragement to tell the story, and for their help in preparing the manuscript

Thanks for the use of clip art from the outstanding artists at Pixabay.com and openclipart.org

Thank you to all my teachers and guides who taught and encouraged me. I'm sure I appreciate your guidance and support more now than I did at the time.

I also appreciate those of you who motivated me by telling me that something couldn't be done, and especially those who doubted me and made it personal by telling me I wouldn't succeed. I love to prove people wrong

ISBN: 9781795772600

© Copyright 2019 by Ed Bernd Jr. and Avlis Productions Inc.

Contents

The Impossible Pictures - 4

Challenges - 6

From Gemini to Apollo - 7

Begging for a Half Hour Off - 11

Improvising - 12

Pushing the Limits - 14

A Nice Day for a Launch - 15

Map of Space Coast area - 17

Mixed Results - 18

A Friend to the Rescue - 19

My Next Project - 21

Apollo 11 - 23

Would We Miss the Moment? - 29

After the Moon Launch Then What? - 31

Not Newsworthy Anymore - 32

From Dawn to Dusk of the Apollo Program - 35

A Night of Indecision - 36

Free Downloads - 40

Pictures for framing - 43 >

The Impossible Pictures

They said it couldn't be done, that I was wasting my time, that it was impossible to take a time exposure photo of a daytime missile launch... in color... directly into a Florida sunrise... from ten miles away.

If anybody would know, Arch Smith would.

Arch had first conceived the idea of making a time exposure of a daytime missile launch during the Gemini program. He ran the photo department at Martin Marietta's Canaveral Division, and he happened to have an opening for a darkroom technician at the same time I found myself between newspaper jobs.

The Gemini rocket used liquid fuel and didn't produce a bright enough exhaust, so Arch photographed the launch of an Atlas Agena rocket that the Gemini crew was going to rendezvous with in orbit.

It would be necessary for Apollo astronauts to rendezvous and dock with the Lunar Lander when they flew to the moon, so Gemini astronauts were proving that it could be done.

Taking a black and white picture of a Saturn 5 launch was easy enough. Even Arch thought I might be able to pull that off despite the fact that the Saturn 5 also used liquid fuel. The Saturn 5 fuel might not have been as bright as the Atlas fuel, but there was going to be a lot more of it.

And I did.

I only had one tripod so I did it with the big Crown Graphic press camera sitting on the hood (or "bonnet" as they call it in England) of my MGA automobile. You can see the bonnet in the bottom right corner of the photo.

Here is a reenactment of that morning. My cab driver shot this picture of me in February of 2019, here in South Texas. I may be a few decades older, living couple of thousand miles from Florida, and my sports car days might be many miles behind me, but the old press cameras haven't changed a bit in the last half century.

The trick for black and white pictures is to use infrared film. It wouldn't work with regular film, but infrared film turns the daytime sky dark, as you can see in the previous photo. Even with the sun shining straight into the lens, it still turned most of the sky dark.

Actually I had already made a similar photo, with Arch watching. I had gone to work at Martin Marietta Company in the Spring of 1966, and Arch showed me the time exposure picture he had taken of an Atlas rocket launch. It was beautiful, but as a news photographer, there was one thing missing:

There was no life in the photo. Not even a houseplant.

So I suggested to Arch that we take another picture and this time have our secretary, Annie, in the shot watching the launch through a pair of binoculars.

The picture turned out great except for one thing: It doesn't look like she is looking at the rocket.

I never intended to have her look at the rocket. If she had been, all you would see is the back of her head. In order to create the illusion that she is looking at the rocket, I had her look slightly to her right, where I expected to rocket to go.

I expected the rocket to go "downrange," towards the southeast where the National Aeronautics and Space Administration (NASA) had established a string of missile tracking stations.

I didn't realize that manned launches didn't go downrange but are launched due east. So Annie is not quite looking at the rocket.

I was a little disappointed, but it paid off later when I made the Apollo photos because I knew where the rocket was going.

Evidently it didn't hurt the picture that she wasn't looking directly at the rocket because the Associated Press photographer who came to cover the Gemini launch from on top of our hanger, put it on the AP wire along with the photos he shot. A couple of weeks later he mailed us a clipping of the photo on the front page of a daily newspaper in the Northeast. We gave the clipping to Annie. It never occurred to me to keep a copy of the picture for myself, and that is why I can't show it to you.

But I can show you this old picture of the Martin Marietta Photo Department in the fall of 1966: Arch, Annie, me, and Bill.

So I was confident that I could get a good black and white photo of the Apollo 4 launch, but could I get a color picture of the same scene?

Challenges

Everything had to go perfectly on the morning of Thursday, November 9, 1967, to keep the Moon Mission on schedule, and to give me a chance to take an impossible picture.

The rocket scientists and engineers had plenty of problems. Just Getting to this day had been even more challenging than they had anticipated.

Back in January, just 10 months earlier, three American astronauts, Gus Grissom, Ed White, and Roger Chaffee, had died in a fire in the Apollo capsule during a routine test.

Everything was shut down at that point as engineers worked to understand exactly what had gone wrong and how to keep it from happening again. The Saturn 5 rocket was the most powerful and complex machine ever built, and resuscitating President John F. Kennedy's moon landing dream in the next 2 years was a daunting task.

Thanks to a series of events that had nothing to do with the Apollo program, I had the opportunity to take a picture of the Saturn1B rocket while it was being prepared for the three astronauts.

The Gemini program was coming to an end after accomplishing everything that NASA had wanted from it.

Gemini is an under-appreciated asset that was created when scientists realized how long it would take to design and build the Saturn 5 rocket.

They turned to Martin Marietta and their Titan II rocket. With two rocket motors and 474,000 pounds of thrust, it could lift a pair of astronauts into earth orbit. Best of all, they could "man-rate" the Titan II within a year – add the redundant systems that are required for manned launches.

By the time the first Saturn 1B rocket was being prepared for launch, Gemini astronauts had already validated most of the procedures that would be required for a trip to the moon:

Gemini was the first to rendezvous with another vehicle in space, and the first to dock with another vehicle in space.

The first EVA – Extra-Vehicular Activity – also known as "space walk" – was by a Gemini astronaut.

Gemini was the first to keep two astronauts in orbit for two weeks. Imagine you and a buddy sitting in the smallest sports car you have ever seen, top up, windows closed, and staying there for the next 14 days. No exercise. No privacy. No bathroom – but you have little bags you can use. When astronauts Frank Borman and Jim Lovell climbed out of the Gemini capsule after their ordeal, the expression on their faces told you know that "once is enough."

From Gemini to Apollo

When the final Gemini flight, Gemini 12, splashed down safely on November 15, 1966, a Saturn 1B rocket was being prepared for the first manned flight in the Apollo program.

Martin Marietta wanted to commemorate Gemini's success with a picture of technicians lowering the erector at Pad 19 for the last time. That erector had lifted every rocket into position for all 12 Gemini missions and everyone involved had become quite fond of it.

Arch sent his photographer, Bill, to take the picture, and I went along too. Martin Marietta Technicians raised the erector about 30 degrees so that we could photograph them lowering it for the last time.

Bill gave me the big Crown Graphic press camera and told me what to photograph and I did, while he climbed onto the gantry and explored various angles and took pictures with our little 35 millimeter Pentax camera.

We came back with great snapshots, but nothing that told the story and conveyed the emotion and pride that everyone associated with Gemini felt.

The public relations director asked Arch to try again. So the technicians went back to Pad 19 and raised the erector again so we could photograph them lowering it for the last time...again.

This time, though, Arch went, and took me along. While he used the big 4x5 press camera mounted on a tripod, he left me on my own with the little Pentax.

Arch covered his head and the camera with a black focusing cloth to shut out light, the way they used to do in the early days of photography. It was quite a sight. He was very popular in the company and everybody wanted to be in his picture.

Here is a picture of Arch with a "Big Bertha" camera, a single lens reflex Graflex camera that used a gigantic telephoto lens to capture distant images onto 4x5 inch film.

Arch left me alone to do whatever I wanted to do, so I did what Bill had done: I climbed onto the gantry and under the gantry, I looked around and over and under everything, and I didn't see anything that would make an interesting picture.

Finally I did what professional photographers do and took a look beyond Pad 19.

That's when I saw it: The picture that would tell the story.

In the background, about four miles away, was Launch Complex 34 and the Saturn 1B rocket that was being prepared for the first manned flight in the Apollo program.

We had a short telephoto lens for the Pentax so I put that on, but none of the technicians would come over to be in my picture. I had to ask Arch to stop taking pictures for a couple of minutes and send a few of the Martin Marietta employees my way so I could have them in the picture.

I fired off about half a dozen shots and sent the people back to Arch.

Arch continued taking pictures and didn't show any signs of stopping, so I walked over and told him – in a kind of arrogant way – that there was no need for him to shoot any more pictures, I had already made the one they were going to use.

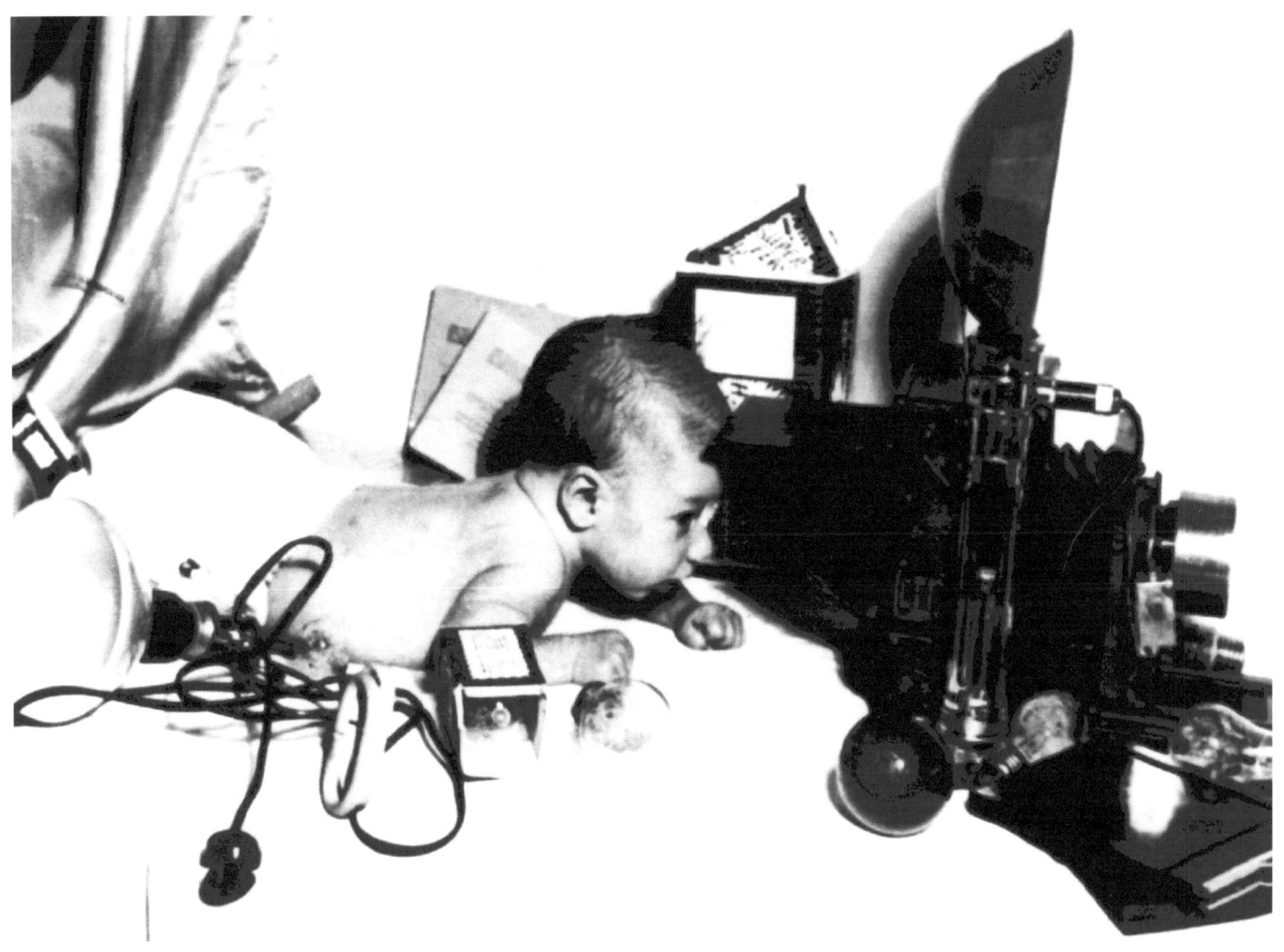

He gave me a hard look, but I knew I could get away with that because we had so much respect for each other. He knew I had been taking news photos all my life, and he also knew how much respect I had for his experience and technical expertise.

If you don't believe I've been taking pictures all my life, just look at this picture of me on our 1940 Christmas Card, when I was six weeks old. The caption was, "Focusing on You."

Arch took a couple more pictures of the Gemini erector, gave me a couple more hard looks – not angry looks, but curious about what I had done that made me so confident – then we packed up our gear and went back to the hanger.

I developed all of his pictures and a couple of mine and laid all the pictures out on a big table – all the photos from both shoots.

Our secretary Annie saw them first, pointed to my picture and said: "That's the one they will use."

When Bill came in he looked at them and as soon as he saw my photo he said, "That's the one they will use."

A few minutes later Arch came in and looked them over. He scanned the photos and when he got to mine, he paused, then he said, "Send them all over to Cocoa Beach and see what they say."

They ran the picture on the front page of the company newsletter, which came out on Friday, January 27, 1967.

That was a memorable day:

It was my last day with Martin Marietta. I got a copy of the newsletter in the morning, and at noon I left for the final time feeling pretty good about my 9 months as working as a Data Acquisition and Analysis Technician in the Gemini and then the Titan III

Engineering Space Programs at Cape Canaveral. I made that title up because it sounds better than telling people my job was developing film and making prints like I had been doing since I was 10 years old.

I headed across the state, picked up my girlfriend and her little brother and we went to St. Petersburg to listen to Lenny Dee and his Hammond Organ. My girlfriend's little brother loved the Hammond Organ.

During the performance Lenny Dee said, in the gentle voice that lounge performers use, "Did you hear what happened at Cape Canaveral today?"

It sounded like he was making a joke.

When he said that three astronauts had died in a fire on the launch pad, I thought: "This is the worst joke I've ever heard!"

As he explained what happened, I realized it wasn't a joke!

My first thought was to cover the story – to reach out to my sources, to talk to people on the street and get their reactions – anything I could do to cover this story.

But who would I do it for? I wasn't working for a newspaper, and every newspaper had its own staff to cover stories like this.

I don't recall anything else about that night. I don't know what we did the rest of the night. And of course the picture was now more a source of sorrow than a source of pride. It was a memorable day okay, but for all the wrong reasons.

Begging for a half-hour off

So now ten months later, after a couple of small tests, NASA was ready to take a giant step forward and launch the massive Saturn 5 rocket for the very first time.

My first challenge in getting a picture of it was getting the okay from my editor at the Melbourne Daily Times to come to work a half hour late.

Pearl Leach is a fabulous editor, exactly the kind of editor you want to work for if you are a police reporter for a small daily newspaper. There is nothing you will encounter that she hasn't seen before, and she will always have your back.

Starting on her 18th birthday, she had worked as a police reporter for a dozen years, and had seen it all.

No matter what kind of jam or embarrassing situation you found yourself in, Peal could empathize with you and share a story from her career that was even more embarrassing.

When you are on deadline and still don't have all the facts, Pearl has been there and done that and you can call on her experience to bail you out.

When the sheriff calls and wakes you up in the middle of the night screaming and demanding a retraction and an apology and promising to be in your editor's office first thing in the morning, you know that Pearl will back you up.

But that's another story – or series of stories – for another book – or series of books.

Hmmm... I wonder if anybody would be interested in those stories. If you are, let me know. If enough people are interested – you know: more than one person – I might write them up.

Meanwhile, back in November of 1967 Pearl wasn't all that enthusiastic about taking an "impossible" picture of an unmanned missile launch, but when she realized how much I wanted to do it, she agreed – provided that I would be in the office by 8 a.m.

I was sure that I could clock in at 8 a.m. – approximately – if the launch went on schedule, and if traffic wasn't too bad. I wasn't too worried though, because Pearl doesn't like mornings any more than I do. I knew that even if I was an hour late, she wouldn't be there to catch me.

Improvising

Now that I had Pearl's okay for a few minutes of time-off, and with Arch still insisting that I was on an impossible mission, it was time for me to figure out exactly how I was going to pull it off.

I had one 4x5 inch press camera, so I borrowed a second one from Sterling Hawk, a local commercial photographer who shot pictures for the Times and processed film for us when we shot our own. We loaded up a couple of cut film holders, one with color negative film, the other with black and white infrared film.

The infrared was easy: a red filter would turn the sky dark, and reduce the amount of light sufficiently for the long exposure. We had done it twice at the Cape: Arch did it first, then we did it again with our secretary Annie in the picture watching the launch.

The color was a new challenge. The film was much faster – I would have to find a way to reduce the light at least 8 f-stops, which I think is 1/256th the original amount.

A 4x neutral density filter only reduces the light by 2 stops, so even if I stacked two of them, it would only give me half of what I needed.

But I had an idea: Polaroid made a special neutral density filter that enabled you to use their ultra-fast 3,000 speed film in daylight. It reduced the light by 4 stops.

Fortunately the diameter of those filters is just big enough to cover the lens on the press camera.

But how to attach them? Polaroid put them into rubber holders that would fit over the lens of their Polaroid cameras, but not over the lens on the Graphic.

So I used tape. I got some black tape and taped one Polaroid 4s filter onto the front of the lens, and another onto the back of the lens.

Now I had two cameras, both set up and ready for action, and film loaded into the cut film holders. But I only had one tripod.

That was no problem: I had often used the hood of my MGA automobile as a camera platform, as well as a big step stool I could stand on if I needed to get up high to get the picture.

So on Thursday morning, I got up early, headed north from my home on South Merritt Island, across the Indian River to US Highway 1, and continued north to Titusville.

Here is a picture of my "camera platform" and me in the fall of 1967.

Pushing the limits

November 9, 1967, was shaping up to be a very special morning.

Long before the tug-of-war began between darkness and daylight, ten thousand people were already up, working, worrying, waiting, wondering if they could really pull it off.

Dawn is not my favorite time of day. Me and mornings don't get along. But this morning was different. With a lot of luck, we would see something that had never happened before.

And I would do something that had never been done before.

A million things had to go perfectly that morning in order for me to get the picture I wanted.

The launch itself was a long shot.

Nineteen sixty-seven had already been a tough year for NASA, and they were improvising to try to fulfill President John F. Kennedy's dream of "landing a man on the moon and returning him safely to the earth" by the end of the decade.

The Apollo 1 fire back in January threw the whole Apollo schedule into chaos. With less than three years left to figure out what had happened, fix it, test the most complex machine ever built and use it to send astronauts to the moon and bring them back again safely, everybody was overworked and overstressed.

After ten months of testing, NASA was ready to launch the very first Saturn 5 rocket.

Gemini's two rocket motors produced a total of 474,000 pounds of thrust.

By comparison, the Saturn 5 rocket, which had more than 3 million parts, produced a thundering 7.6 million pounds of thrust at liftoff.

That was so much power that they could not start all five motors at once. Some people feared it would knock the earth off its axis. Engineers feared it might blow the rocket apart if they ignited all 5 motors at once. So they started igniting the 5 rocket motors 3 seconds before liftoff.

To paraphrase one of the astronauts, imagine sitting on top of thousands of pounds of high explosives in a vehicle built with 3 million parts... all provided by low bidders on a government contract.

So on that November morning, everything had to go perfectly to keep the Moon Mission on schedule... and to give me a chance to take an impossible photo.

I loaded everything into the MG and started the 45 mile drive from my home on South Merritt Island to Titusville.

A nice day for a launch

There is a nice area on the bank of the Indian River at the intersection of US Highway 1 and State Road 50 at Titusville where you can pull onto the shoulder and look across the Intracoastal Waterway and see the largest building in the world, the Vehicle Assembly Building (VAB). It is not the tallest building, but has the most volume. It is clearly visible in my photos, even from more than ten miles away.

After the Apollo program ended, a lot of people watched Space Shuttle launches from that spot, and those launches were just as spectacular as the mighty Saturn 5.

I found a nice spot for the MG, close to a little palmetto bush that would look nice in the foreground to give some depth to the picture.

A lot of people had come to watch the early-morning launch, including an excited and nervous young woman whose husband worked on the rocket. She was listening to the countdown on a portable radio. I set up my cameras next to her so I could hear the countdown too.

Everything was going exactly as I had hoped as we listened to the countdown. The edge of the sun crept over the horizon at about T-minus-20 minutes before liftoff.

They had to "light the candle" on time – without any delays – in order for me to get the picture I wanted. Or any picture at all for that matter – I couldn't wait for a launch later in the day.

I watched and waited for red and orange and yellow color to fill the horizon and sure enough it did. The color spread and intensified as the countdown continued.

Suddenly my opportunity to shoot my impossible photo and make my "impossible dream" come true began to disappear right before my eyes.

The sunrise began fading away.

There had been a beautiful red glow about 5 minutes before the scheduled 7 a.m. liftoff. And then it faded away.

I watched helplessly as the color kept slipping away. It looked like it would be completely gone about two minutes before liftoff. There was nothing I could do about it, nothing I could do to overcome this challenge.

Sure enough, with two minutes to go until liftoff, all the color was gone.

Then a miracle happened:

As the sun continued to climb up over the horizon, it found a little white cloud just a few degrees above the horizon.

As the count closed in on the final seconds before liftoff, the sun slowly slipped behind that little cloud, and the color began to return.

I held my breath as more and more color spread out behind that skinny little cloud, not as much as the "real" sunrise, but enough for my picture.

As the count reached T-minus-3 seconds and they started igniting each of the five engines, I opened the lenses on both cameras. The color started fading at about that time, but there was more than enough color still left during the next 60 seconds for me to make my picture.

As the count reached Zero and the giant rocket began to slowly power its way up – right on schedule – people started yelling and cheering it on. The young woman next to me was jumping up and down and shouting enthusiastically, then she grabbed my arm and held on enthusiastically.

It felt good, but not good enough to make me forget about my picture. I reminded her to please be careful not to bump the tripod or the car. She quickly turned loose, moved a couple of steps away, and continued jumping and cheering. Then I wished I hadn't said anything to her. Excited young ladies excite me. But to me, the job always comes first. (sigh)

After 60 seconds I closed the shutters on both cameras, loaded everything into the MG, and began my next challenge: To get back to the Times office in Melbourne almost 60 miles away by 8 a.m. I'd promised Pearl that I could do it.

To give you an idea of how far I had to go, here is a map of the Space Coast – or the "Platinum Coast" as newscaster David Brinkley called it one night in a sarcastic tone of voice. He said if south Florida was the Gold Coast, then Brevard County must be the

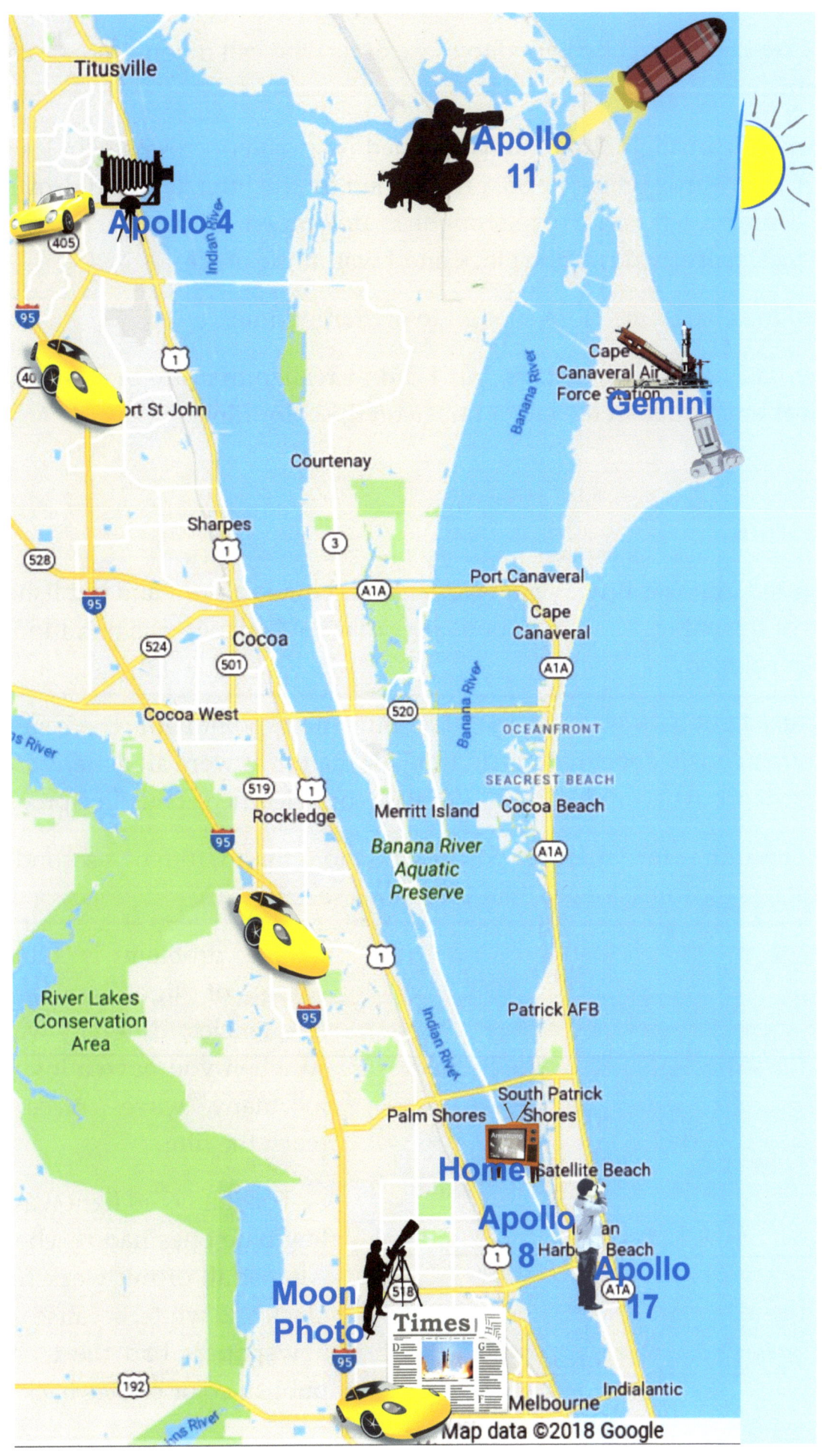

Platinum Coast. That sounded good to those of us who call it home, so we continued to use that term.

There was surprisingly little traffic, so I had no problem crossing US 1 and heading west on SR 50 towards Interstate 95. I turned left on the Interstate and drove until I got to U.S. 192, took a left and drove straight to downtown Melbourne. A right turn on Waverly Place, halfway down the block and I was at the office.

I clocked in at 8:04 a.m., close enough to my target time.

The day had been a big success, but I didn't realize that my biggest challenge lay ahead. It was embedded in the latent image on the color film.

Mixed results

The infrared film was fine. We developed it and made a print and published it on the front page of the paper. That afternoon, after that day's paper was out, I took the color film to be developed.

The picture I got back was a shock and a huge disappointment. The beautiful colors that had miraculously appeared just in time for the liftoff were all gone. Vanished. The whole picture had an ugly reddish color. That is when I learned about "reciprocity failure."

Ektacolor S film is for "short" exposures, no more than 1/10th of a second. Ektacolor L is for "long" exposures, longer than 1/10th of a second.

The problem is that different colors of light are at different frequencies, different wavelengths, so when you have a long exposure, too many "waves" of certain colors reach the film.

Too many red light waves and too few blue ones had reached the film. Almost all of my work had been in black and white, because virtually no newspapers had the technology to publish color photos back then.

I sent the negative to a couple of different photo processing labs and the prints came back the same.

A friend to the rescue

Then a photographer friend that I had met when he was a member of the city council told me about Color Lab of Florida in Ft. Lauderdale. I sent them the film with an explanation of what I had done, and waited to see if they could help.

They helped – they rescued my "impossible mission."

They sent the whole series of test prints they had made. The first was the same as I had gotten from the other labs.

The second was better, but not great.

The third one was good enough. There was some blue in the sky, contrasting with the yellow and orange in the sunrise. It looks like the sunrise is trying to fill the whole sky. That was good enough for me. We were both very impressed with each other, and I continued to send all my color work to them for the next 25 years.

Then 50 years later when I decided it was time to do something with my Apollo Launch Photos I bought a couple of scanners – an old Epson flatbed scanner capable of scanning 4x5 film, and a CanoScan 9000F – and began experimenting.

The scanners are smarter than I am! The software for both scanners have a couple of marvelous adjustments: One adjustment lets you compensate for fading colors, the other is for "backlight correction." Both offer three levels: light, medium, and heavy correction. So I tried all of the combinations to see what would happen.

Those adjustments fixed the photo! When I tried to see what I could do manually, it was never as good as what the software had done on its own.

Finally I have a photo that looks like the scene looked that morning more than 50 years ago.

My next project

The next year, 1968, was a rough one in the United States.

Robert Kennedy and Martin Luther King Jr. had both been murdered.

There were riots and major American cities were on fire, literally.

Every day newspapers carried the names of young American soldiers who were being killed in the war in Vietnam.

People were marching outside the White House chanting, "Hey, hey, LBJ, how many kids did you kill today."

Before the end of the year, President Lyndon B. Johnson announced that he'd had enough, he was quitting, and would not seek reelection.

The whole country was angry, frightened, and depressed.

As we neared the end of 1968 NASA realized that they only had one year left to meet President Kennedy's challenge to "land a man on the moon and return him safely to earth by the end of the decade."

So they decided to double up on their tests and they scheduled the scariest launch yet. They would use it to run two major tests, with live astronauts on board:

They would put 3 men on top of the rocket, strapped securely inside an Apollo capsule like the one three astronauts had died in the previous year.

Then, instead of just launching them into orbit, circling the earth a few times and bringing them back down, they would send the to the moon.

Not to land on the moon, just to circle the moon and come back home. Hopefully.

Once they got the schedule worked out, they realized that the astronauts would arrive at the moon on Christmas day.

That could make it the greatest Christmas celebration of our lifetime, or a Christmas that we would all want to forget.

It wasn't until many years later that NASA confirmed what we all suspected:

They figured the odds that Apollo 8 would make it back to earth safely and that the crew would survive were no better than 50-50.

What we didn't know at the time is that NASA asked the astronaut's family members what they thought, and they were unanimous:

It is worth the risk – let's go to the moon!

Naturally that was a picture I had to take.

So did Arch.

Arch decided if I could do it, so could he. He made arrangements to take a daylight time exposure of the Apollo 8 launch – in color – from on top of a tall building on the beach in Cocoa Beach.

The night before the launch, he called me, worried:

"It won't work!"

I laughed and reminded him that it already had worked. "You saw the picture!"

"But you can't reduce the light enough with two 4x neutral density filters," he said. "Each filter only reduces the light by 2 f-stops."

"You are right about that," I agreed.

"So it wont work!"

"We aren't using 4x filters," I reminded him. "We are using Polaroid 4s filters. The 'S' stands for 'stop.' Each Polaroid filter reduces the light by 4 f-stops."

So while I went down to the Mather's Bridge fish camp at the southern tip of Merritt Island the next morning, Arch went to the top of the tall building in Cocoa Beach. He insisted that he got a good picture. I asked him to show it to me. Through the years I asked him several times to show it to me. He never did.

This time I used Ektacolor L film and the colors came out perfectly. That goes to show that I can learn from my mistakes.

Apollo 11

There are plenty of photos of the Apollo 11 launch, and many of them are higher image quality than mine. If I had been using a Nikon instead of a Pentax, then Nikon would have loaned me a big lens for the shoot. They were happy to loan equipment to press photographers, because then they could brag about all the moon launch photos that were made with Nikon gear.

But I was on my own, and I wanted to do something different.

Arch Smith had a theory that a good amateur photographer could always take better pictures than a professional.

I was a good example of that on this assignment – mainly because it wasn't an "assignment." I did not have to come back with a photo, so I was free to experiment and take chances. If I came back empty handed, nobody would know but me.

My first challenge was coming up with an idea that I hadn't seen before. An obvious idea would be to include both the Apollo 11 rocket and the moon in the same picture. And to do it in-camera, not in the dark room.

Even today when I do an Internet search for "photo of apollo 11 rocket and the moon" and "photo of apollo 11 launch and the moon" I don't find any pictures like that. So maybe my double-exposure photos are the only ones in existence.

However, Life magazine had a similar idea: They took a picture of the three astronauts ahead of time, and then double exposed that with the rocket launch. I think they might have done that for both Apollo 11 and Apollo 14, because I met a very nice Life magazine photographer during the Apollo 14 launch who was taking pictures with about a dozen cameras at once, including the multiple exposures in both black and white and color.

The first two problems I needed to solve were where to find lenses suitable for my purpose.

Sterling Hawk at Sterling's Photography had the solution for a lens that was long enough to photograph the liftoff from 2 miles away. That's how far it was from the launch pad to the very front edge of the press site.

Sterling had a 500mm mirror lens. By using mirrors, like with a telescope, the lens would be far less than 500mm long. It would be easier to handle than a longer lens. But it did lose some sharpness. Regardless, it was the only option I had, and it would definitely do the job. The image of the rocket is 3/4 of an inch tall (18mm) on the 1-inch-tall 35mm film.

But Sterling didn't have a lens long enough to give me an image of the moon that was at least 3/4 inch tall.

While talking about it, someone told me they knew someone who had a brother who had a big telescope in his back yard, so I tracked him down and he was kind enough to let me use his telescope to photograph the moon a couple of weeks before the launch.

I marked the film before I loaded it into the camera so that I could take it out, and then reload it exactly the same so I could make the second exposure on the same frames two weeks later.

But how do you take a picture through a telescope?

I had time to think about it as I stood in his backyard watching and waiting for the moon to rise above the trees so I could get a clear shot of it.

I got a little nervous as a police car drove down the street. I was a police reporter so I knew most of the officers, but what if it was an officer who didn't like something I had written about him or the department. But he kept going, so I went back to my moon gazing.

The telescope was huge, and was mounted on a pedestal. It had a mirror that reflected the light back and gave it a focal length far greater than the length of the physical barrel. It had a removable eyepiece in the side of the barrel.

I put the lens of the Pentax against the eyepiece and took a look through the viewfinder. That didn't work.

I took the lens off the camera and tried using the eyepiece as the lens. I don't think that worked either. Well, that was 50 years ago and it was the only time I ever took a picture through a telescope, so forgive me if my memory is a little fuzzy.

I think I ended up removing the eyepiece – partially dismantling the man's telescope – and just holding the camera up to the hole in the side of the telescope and letting the mirrors reflect the image of the moon onto the film. It looked good in the viewfinder.

With that problem solved, I started working on the next challenge: What exposure do you use when taking a picture through a telescope?

Fortunately the Pentax had a nice through-the-lens exposure meter.

Unfortunately, the needle that indicated how much light was going to the film was on the edge of the viewfinder. The was a problem because the only light was the round image of the moon itself in the center of the viewfinder. It didn't illuminate the exposure needle.

I could move the camera so that the moon was on the edge of the frame and lit the area where the needle was, but that didn't give me an accurate reading: The exposure meter is "center-weighted," which means it got most of the reading from the center of the frame.

I could see the needle dropping as I moved the camera towards where I could see the needle.

In other words, I couldn't get an accurate reading.

So I guessed.

That wasn't the only thing I guessed at.

I had guessed that it would be better to take the picture either before or after the full moon. That way, I reasoned, the slight shadow on the edge of the moon would provide a little depth – some dimension – instead of just a bright disk. That seems to have worked.

I skipped the first couple of frames, so that I would be able to take pictures of the rocket alone as it began to lift off the launch pad. It took almost eight seconds for the rocket to clear the launch tower, so that gave me plenty of time to expose a couple of frames.

After it cleared the tower, then there would only be blue sky behind the rocket, the conditions I needed for the double-exposures with the moon.

I was also guessing what it would be like to take the pictures. I was using a lens that was built like a telescope to take pictures of the launch. I had never used a telephoto lens that long, and I had never even been to the press site at Kennedy Space Center for a launch, much less taken a picture of a launch from just two miles away.

I alternated, exposing a couple of frames of the moon, then leaving a couple of blank frames that would only show the rocket. Then more frames of the moon, followed by more frames with the moon.

Boy oh boy was I in for some surprises.

First was the sound. The rocket is big enough to shake the ground 30 miles away in Melbourne. It is a deep rumbling sound.

Across the river on U.S. 1 it is the same kind of sound, but louder.

It is not that way at the press site: It sounds and feels like an explosion! Not a problem though. I was a little startled by the sudden intense sound, but I got my shots off.

Then after waiting patiently for the rocket to clear the launch tower, I thought I would have plenty of time to expose the next 34 frames.

Not so. Once that big bird gets moving, it flies fast. Very fast.

I had the camera on a tripod, and had the tripod adjusted to its lowest setting. Things tend to look bigger when you shoot from a low angle, close to the ground. That is great when you are photographing athletes from the sidelines, but when you are taking pictures of a giant rocket from 2 miles away, it doesn't make any difference in the picture.

But it sure made a difference in shooting the pictures. The rocket began to accelerate so fast that I could barely keep up with it. I was snapping pictures and winding the film as fast as I could, and kept rotating the camera back on the tripod, pointing it higher and higher until I felt like I was about to fall on my back.

Before I had finished the roll of film, I knew the rocket was too far away for the pictures to be of any value, but I kept shooting anyway.

The pictures came out the way I expected. Nothing spectacular, but I had photographs of the most historic event of my lifetime.

I was just one of what seemed like a million photographers at the press site that morning. I knew my photos of the rocket blasting off the launch pad wouldn't be anything special, but my double-exposure photos with the moon seem to be unique.

Here is a NASA photo showing part of the scene at the press site that day. I was part of the crowd way up at the very front, as close as they would let us get to the rocket. In this photo the rocket is just clearing the launch tower, about eight seconds into the flight. Another 10 seconds or so later it was like craning your neck to look up at the ceiling.

Now that I had my pictures of the launch, I was ready to take the picture I really wanted: An astronaut stepping onto the surface of the moon.

Would we miss the moment?

Apollo 11 arrived at the moon right on schedule. Neil Armstrong and Buzz Aldrin climbed into the Lunar Lander while Michael Collins waited behind in the Apollo 11 spacecraft.

Then there was the exciting descent to the surface of the moon, when Armstrong had to take over control and fly the Lander manually in order to maneuver to a smooth landing spot.

Nobody on earth knew what was happening during those long moments, until Armstrong got back on the radio and said those famous words: "Tranquility Base here, the Eagle has landed."

It was the perfect phrase: The first words – "Tranquility Base" – meant they had landed. The rest of the phrase was confirmation.

Then they began their preparation to open the hatch and climb down the ladder to the surface of the moon.

A small group of us were poised around my old twenty dollar black and white television set at the same time, making our preparations. I had the 4x5 press camera on a tripod, loaded with Polaroid 4x5 professional sheet film, and a young boy sitting in front of the camera, watching the television.

Finally the big moment arrived. Neil Armstrong had emerged through the door of the Lunar Lander and was standing on the top step, just moments away from taking the first historic step onto the surface of the moon.

We were watching and waiting to see the most historic event of our lifetime – perhaps of all time – when we heard a screech! and a crash! outside.

Some idiot had driven down South Tropical Trail too fast and wrecked his car in my front yard! It happened occasionally. There is a little jog in front of the house that catches drivers by surprise and they hit one of the palm trees.

My best friend Bill was there. He is a first responder, a volunteer fire fighter and ambulance attendant, and I have seen him risk his own life to go to the aid of others.

But this time we sat and looked at each other, undecided about whether to stay and photograph history, or go check on the idiot in my front yard. If he is that dumb, we wondered, to get drunk and wreck his car at a moment like this, then are we actually doing the world a favor by saving his life?

But of course we knew we had to go check on him.

As we started to get up we heard a woomp! woomp! woomp! and chug! chug! chug! sounds of an automobile engine turning over, then it caught and started to run. The chug! chug! chug! was accompanied by the clank! clank! clank! of what sounded like a fan blade hitting metal, probably the radiator.

So Bill and I waited to see what would happen next.

Sure enough, we heard a ker-klunk! as he put it into gear, and then some scraping and squealing and clanking and thumping sounds as the car pulled away and continued down South Tropical Trail. Away from my front yard.

Now that the driver was gone and there was nothing we could do for him, we sat down and turned our attention back to the television set.

I got behind the camera with my finger on the shutter and when Armstrong jumped onto the surface of the moon and they flashed the words "Armstrong on Moon" I tripped the shutter and we had captured history on a sheet of Polaroid film.

I had traveled up to Kennedy Space Center – which was originally called MILA – Merritt Island Launch Area – at the north end of Merritt Island, and had taken a picture of the rocket as it began its journey to the moon.

Now I was in my home on the south end of that very same island – Merritt Island – taking a picture of a young boy watching as a human being stepped onto the lunar surface for the very first time.

It was very satisfying.

After the moon launch, then what?

Things were strange after the Apollo 11 launch. It actually began before the launch, when aerospace companies started laying off people. We had accomplished our mission, so the rocket scientists were no longer needed.

The question is: What does a "rocket scientist" do when there is no more need for rocket scientists?

All of their education and training, and all of their work experience, was in rocket science and engineering.

A lot of them opened restaurants. Everybody has to eat, right? So surely that is a good business.

It is also a difficult business, especially if there is a lot of competition.

And there was a lot of competition:

The director of the Brevard County Health Department told me that they had more requests for inspections of new restaurants in the year after the moon landing than they'd had in the entire history of the Brevard County Health Department before that. Most of those restaurants failed.

Others bought service stations, but they had no experience or training in running a small business, and most of them soon gave it up.

There were a lot of suicides, which we seldom covered in the newspaper.

For years Brevard County had been one of the most stressful places in the United States, with more alcoholism and a higher divorce rate than anywhere else in the nation.

The aerospace industry contributed to the stress: To save money, the aerospace companies would lay people off just before they were eligible for benefits.

That happened to me when the Gemini program came to an end. I had worked for Martin Marietta Corporation for nine months, and during that time – in addition to being paid more than my work was worth – I had accumulated almost a full month of vacation time and another month of sick leave. I didn't have another newspaper job yet – I wanted to stay in Brevard County – and I figured I would have a couple of months extra pay to tide me over.

That didn't happen. Arch explained that in order to be "vested" I had to have been employed there for a full year.

A lot of people already knew about that: They would work for an aerospace company for almost a year, then get laid off and hired by another company a few weeks later. No vacation pay, no compensation for unused sick leave, no retirement benefits. Just the stress of looking for and then starting a new job every few months.

Not newsworthy any more

The whole world watched on television as Neil Armstrong stepped onto the surface of the moon. To me, it is still the greatest achievement in the history of the world.

But after that, people lost interest in the space program. We had already been to the moon, and as the old saying goes: Nothing is older than yesterday's news.

While reporters and photographers and dignitaries from all over the world attended the Apollo 11 launch, nobody came to watch Apollo 12 take off on its visit to the moon.

Well, not "nobody," but very few came to cover it, and it received very little press.

So prior to the Apollo 13 launch, NASA got our attention by launching an Air Force rocket at night without announcing it ahead of time.

Nothing will get a reporter's attention faster than acting like you are trying to keep a secret – to do something without letting the news media find out.

So the Apollo 13 launch was better attended than Apollo 12, but far short of Apollo 11.

And then there was Apollo 13's well documented "problem" that almost killed the three astronauts. If you are not familiar with it, then watch the outstanding Apollo 13 movie.

That was not a manufactured problem, it was real, and it reminded everybody that space flight was serious business and we needed to pay attention. There was no shortage of news coverage after that.

We had new editors at the Melbourne Times and they decided we should all go to Kennedy Space Center for the launch of Apollo 14. Our assignment:

Get a story that nobody else has.

Sure, nothing to it: Find a story that a thousand other hungry reporters missed.

I took a black and white picture when the countdown clock reached zero.

There was nothing exciting about that picture, but it brought some symmetry to my Apollo photos. I had that tragic photo of Apollo 1 on the launch pad and then the launch of Apollo 4. Then picking up at Apollo 8, I photographed every third launch: 8, 11, 14, and finally 17.

While at Kennedy Space Center for the Apollo 14 launch I met a Life magazine photographer who was using a dozen or so cameras to take a whole series of pictures all by himself.

As I recall, he had 3 tripods, each with multiple cameras attached to a horizontal bar that was mounted onto the top of the tripod. He also had three foot switches to so he could trip the shutters remotely, and he had another camera mounted on a NASA photo platform that was there to shoot film of the rocket when it was launched for as long as it was in sight. There was a tracking sensor on the platform to keep the camera centered on the rocket's exhaust. If it blew up, they would have pictures of it.

The Life photographer had done what I did for Apollo 11 and exposed some film ahead of time, only he took pictures of the 3 astronauts: Alan Shepard, Stuart Roosa, and Ed Mitchell. He had done them in both black and white and in color, and had cameras loaded and ready to superimpose the rocket launch onto the film after the rocket had cleared the launch tower. Those cameras were on one tripod so he could start taking pictures with them after the rocket cleared the launch tower.

He had other cameras, some with black and white film and others with color, to shoot tight shots of the rocket, wide shots, and who knows what else, as they left the launch pad.

What I found interesting was a sketch artist who noticed the camera setup, and sat down on the ground and started drawing it on his sketch pad.

I found an angle where I could have the sketch artist, the Life cameras, and the rocket that the cameras were pointed at all in the photo.

There were a lot of people at the press site, walking around, waiting for the launch, so I had to wait until nobody was in the way and then take my picture. I've been doing this for a long time, so I am accustomed to it. When necessary, I keep both eyes open so I can see what is going on outside of the viewfinder.

Eventually the crowd cleared and the only person I saw off to the side was the young Life magazine photographer walking back towards his cameras, so I went ahead and tripped the shutter.

The Life photographer heard it – we photographers are very familiar with that sound and it gets our attention. As soon as he heard it, he started apologizing to me:

"I'm so sorry, I didn't mean to step in front of you," he said, and I replied, "You didn't, I had both eyes open, I saw you coming and I took the picture before you got into the frame."

"I'm sorry," he said again, "I'd never do that to another photographer." And I replied again, "You didn't, I saw you coming. I had both eyes open."

I think I finally convinced him. He was a very nice guy. Life magazine photographers were the best news photographers in the world. I was impressed that he was so humble and so concerned about somebody he didn't even know.

I wonder if that episode qualifies as a story that nobody else had. Perhaps, but nobody would be interested except other photographers.

It was an interesting day, but other than my brief encounter with Life magazine – which is always fun for an old news photog – the day was pretty uneventful.

From Dawn to Dusk of the Apollo Program

The mighty Saturn 5 rocket had first roared to life at dawn on a on a bright and clear Thursday morning in 1967. Now it was scheduled to blast off on its final journey into space five years later on a damp and cloudy Wednesday night in December of 1972.

It should have been the easiest picture to take, but it turned out to be very eventful. After the shoot I was disappointed and depressed and wondering how I could have done so poorly.

But when I finally got the picture developed a few weeks later I was dumbfounded and ecstatic and wondering how I ever snapped such a spectacular picture.

In the end, this final Apollo launch photo turned out pretty good, to say the least.

I was working for a different newspaper. In the summer of 1971 the Orlando Sentinel had hired me to be their Melbourne bureau chief. They had specialists who covered the space program, and they weren't interested in anything I had to say about it. So I sat out the Apollo 15 and 16 launches and spent my spare time covering the local stock car races.

But Apollo 17 was different: It was the last manned flight to the moon, my last chance to photograph a manned moon launch, and it was the only Apollo mission to take off at night. I couldn't pass all that up.

It didn't pose the same problem as the daytime launches of finding a way to reduce the amount of light painting the film, but I still had the challenge of what exposure to use. As far as I know, nobody had ever done this before, so all I could do was make an educated guess. And my education in this area was pretty limited.

My previous experiences had taught me exactly where the launch would come from and exactly what direction the rocket would go. The experience I'd had more than 6 years earlier when I posed our secretary Annie too far to the right for my first time exposure missile launch photo turned out to be a valuable lesson. It helped me get the composition right for the Apollo 8 launch photo from the old Mather's Bridge fish camp, and the additional experience of that photo gave me confidence that I knew exactly where to point the camera for this final moon shot.

But I still needed something else in the picture besides the streak from the rocket's exhaust.

I decided to shoot the picture from the remains of the old Canova Pier that jutted out into the Atlantic Ocean just off the east end of the Eau Gallie Causeway. The contrast of a relic of early 20th century Florida with the technology of a moon launch would make a nice contrast.

All that I would need to do is to set up my cameras at the right distance from the remains of the old pier, and then figure out how to get enough light on the pier so that it would be visible against the dark sky.

A night of indecision

Those two things turned out to be much more challenging than I expected.

There were several delays in the countdown, with rain accompanying each delay.

It seemed like Apollo was desperately clinging to life, realizing that once they lit his fuse, he would take flight for the very last time, and would never fly again.

Those delays and the rain were a blessing as far as the composition was concerned:

Each time it began to rain, I took my cameras underneath the pier to keep them – and me – dry.

Then when the rain stopped I came back out, walked back to my previous spot south of the Pier – and decided I needed to back off even farther.

I did that same dance several times, each time moving farther and farther back from the pier.

Judging from the results, I guess I finally got it right.

I only had a small flashgun with me. My plan was to walk up close to the pier, after I had opened the shutters, and fire the flash at the pier as many times as I could during the 60 second exposure. People had gathered north of the pier to watch the launch, but I thought they were too far away for the light to reach them.

I turned out to be wrong on both counts.

Eventually I realized that the old weathered wood of the pier would soak up all the light from my little flash unit, and it did.

But my "open flash" techniques wasn't wasted, because it lit up the people on the beach and that added a lot to the final photo.

I was using 2 cameras, both with color film. One had color negative film, while the other camera was loaded with slide (transparency) film. My plan was to put the slide film into the overnight pouch to Orlando with a note asking them to develop it and see if they wanted to publish it in the Orlando Sentinel.

But after the launch, I abandoned the idea. I was still wondering if I had ever found the right spot to shoot the pictures from, and after firing the flash at the pier I was very doubtful that it had lit up the pier enough to make a decent photo.

No point in sending the editors at the Sentinel a photo of a black sky with a curved streak going through it.

My decision not to send the film to Orlando turned out to be a good one: The frame was blank! Other frames on the roll of film were exposed. You could see traces of light from lanterns that people carried across the beach. But there was nothing at all on the frame that should have had the missile launch.

Only one of my cameras had a "T" setting for Time Exposure, where the shutter would remain open until you pressed the shutter release a second time. The camera that I used for the slide film only had "B" for Bulb, where the shutter remained open for as long as you held the shutter release down.

Since I couldn't stand there and hold the shutter release during the launch, I got a roll of tape, wrapped some around the shutter release, and then let the weight of the roll hold the shutter open. The practice frames showed the trails of light from the lanterns that people had carried, and proved that my idea was good. It just didn't work when it counted.

Now my only hope of having a picture was with the lone remaining roll of film from the other camera.

The Identification Officer at the Melbourne Police Department had taken a similar time exposure picture of the launch, but farther north at the Pineda Causeway. The composition was fine, but the color was horrible – an ugly shade of yellow. We never did figure out why. Maybe because he was behind the rocket and it was going away from him, perhaps the exhaust overwhelmed the film. Whatever happened, it sure didn't make me feel any better about my own shot.

Everything seemed to be going wrong. I wasn't very optimistic. In fact I was disappointed, discouraged, and concerned that I had blown my opportunity through poor decisions and bad luck.

It was already mid-December, two weeks before Christmas and the Postal Service was airing their annual plea to mail your packages early, and I decided not to risk having the film get lost in the mail.

So I didn't even mail it to Color Lab of Florida until January. Perhaps I was a bit like Apollo on that cold damn night, clinging desperately to my last hope of getting just one more Apollo launch photo.

When I finally mailed the film to Color Lab of Florida, I included a note to them explaining what I had done and my hope that the flash might have provided a little bit of light on the pier so that it would show up against the night sky.

When I picked up the package at the post office a week later I was still in no hurry to see the results, so I didn't even open it until after I went outside and sat down in the MG.

When I opened the box there was a note on top apologizing and telling me they had tried but were unable to get the pier to show up.

Then I pulled the 8 by 10 inch print out of the manila envelope and was stunned at what I saw:

It didn't look anything like I had expected. It didn't look anything like I had imagined. It didn't look anything like any other missile picture ever made.

It was – and still is – the most beautiful missile launch photo I have ever seen.

The pier was exactly where I had wanted it to be. The missile went exactly where I anticipated it would. The pier is far more visible than I ever imagined… as a silhouette. And while the flash didn't light up the pier, it did light up dozens of people lined up on the beach to watch the launch.

All I can say is thank goodness I was sitting down when I first saw it, because if not, I think I would have fallen down.

Many people consider the Apollo 17 launch photo the best I ever made. It is definitely the most popular, which illustrates what Arch Smith said: that a good amateur could take better pictures than a professional at any time.

I had a dozen or so copies framed and gave them to some of the public officials and others who had provided me with numerous news stories through the years. It was fun going into city and county offices and seeing my Apollo 17 picture on the wall.

I had given a print to Bill and Barb Cramblit at Bill's Bait and Tackle and they put it in a lighted beer sign. It was beautiful, but I new that a transparency would look even more spectacular, so I had one made for them and they displayed it for a long time.

Any fool who had a box camera loaded with long exposure color film could have captured that photo, as long as there was a way to keep the shutter open for 60 seconds.

But let's keep that between us. I'm quite happy to take credit for it. After all, as somebody said:

The harder I work, the luckier I get.

Download Pictures for Free

The Apollo launch photos look great on a digital screen, like those Digital Photo Frames you can buy, so I made copies of all of the missile launch photos at 1024 by 768 pixels, the 4x3 proportion that looks best. You can download them for free from www.ApolloLaunchPhotos.com

There is also a link on the web site where you can order photographic quality prints, canvas prints, and more. You can also order custom frames if you wish.

There is an example below of one of the frames. It is very similar to the ones I gave to public officials to hang in their offices and homes.

And on the pages that follow are full-page prints that you can cut out display.

I hope you enjoy these pictures as much as I have. I wasn't assigned to shoot them, I didn't shoot them for pay, and evidently I wasn't very concerned about getting them published since it has taken 50 years to get around to it.

This is the first time that the color versions of these photos have been published.

I also hope you remember Arch Smith's lesson, that an enthusiastic amateur can do better work than a professional at any time.

Thank you.

For more information, for free pictures you can download, for videos you can watch, and if you'd like to buy photographic-quality prints with an option for custom framing, please visit www.ApolloLaunchPhotos.com

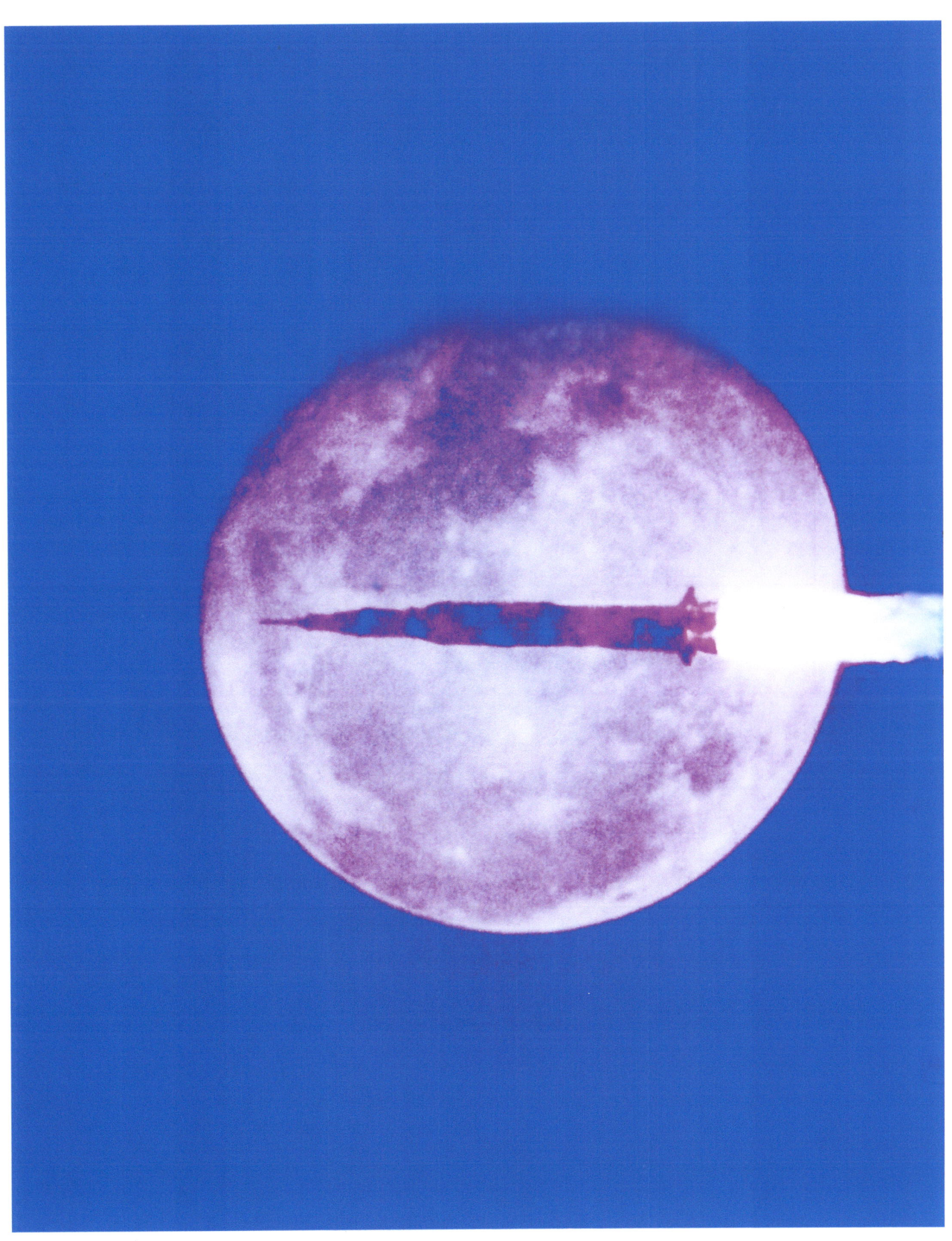

www.ingramcontent.com/pod-product-compliance
Lightning Source LLC
Chambersburg PA
CBHW051215220526
45473CB00003B/1043